Alex Moulton

A LIFETIME
IN
ENGINEERING

An Interview
with
Alex Moulton and John Pinkerton

LIT

Edited transcript of the video programme by Tony Hadland
and John Pinkerton. Autumn 1998.
Interview © Tony Hadland and the estate of the late John Pinkerton

Bibliographic information published by the Deutsche Nationalbibliothek
The Deutsche Nationalbibliothek lists this publication in the Deutsche
Nationalbibliografie; detailed bibliographic data are available in the Internet at
http://dnb.d-nb.de.

ISBN 3-8258-0755-9

A catalogue record for this book is available from the British Library

© LIT VERLAG Dr. W. Hopf Berlin 2007

Auslieferung/Verlagskontakt:
Fresnostr. 2 48159 Münster
Tel. +49 (0)251–62 03 20 Fax +49 (0)251–23 19 72
e-Mail: lit@lit-verlag.de http://www.lit-verlag.de

Distributed in the UK by: Global Book Marketing, 99B Wallis Rd, London, E9 5LN
Phone: +44 (0) 20 8533 5800 – Fax: +44 (0) 1600 775 663
http://www.centralbooks.co.uk/acatalog/search.html

Distributed in North America by:

Transaction Publishers
New Brunswick (U.S.A.) and London (U.K.)

Transaction Publishers
Rutgers University
35 Berrue Circle
Piscataway, NJ 08854

Phone: +1 (732) 445 - 2280
Fax: + 1 (732) 445 - 3138
for orders (U. S. only):
toll free (888) 999 - 6778
e-mail:
orders@transactionspub.com

INTRODUCTION

About a hundred miles west of London and not far from the historic city of Bath, lies the little town of Bradford-on-Avon. Situated in the county of Wiltshire, it is close to the border with Somerset and on the southernmost edge of the Cotswold Hills. Bradford means 'broad ford', the name dating back some 1300 years to Saxon times. That was more than 500 years before the first bridge was built across the Wiltshire Avon.

It was not until the 13th century that a narrow packhorse bridge was erected, part of which is incorporated into the present town bridge. The Saxons built Bradford's most famous building: the tiny church of St. Laurence, rediscovered in the mid-1800s. Because it had been converted to secular use and hemmed in by other buildings, its existence as a chapel had been completely forgotten.

For more than 600 years, wool was the staple industry of Bradford. Some former wool mills still stand on the banks of the Avon, which provided water power to drive the machinery. The town also had a quarrying and masonry industry. The local area provided durable Bath stone for walls and more easily split stone for roofing slates.

However, by the 1840s the traditional industries were in deep decline, and Bradford-on-Avon was far from prosperous. It was then that a newcomer arrived in Bradford, a man born in Devon of seafaring stock who had become a broker and who for some years had been working in New York. His name was Stephen Moulton.

While working in the USA, Moulton had met Charles Goodyear, who had discovered how to convert natural rubber into a strong, yet elastic material by heating it at high temperatures with sulphur. Goodyear called this process 'vulcanisa-

tion'. Stephen Moulton was quick to appreciate the possibilities.

He obtained the British rights to Goodyear's process and returned to England, where he tried to interest the government in the potential of vulcanised rubber. Whitehall was not interested, so Moulton decided to manufacture rubber himself. At the suggestion of one of his financial backers, Stephen Moulton moved to Bradford-on-Avon, where in 1846 he acquired the relatively new but disused Kingston cloth mill.

Two years later he started rubber production and was soon supplying products ranging from rubber buffer springs for the railways to waterproofs for military use. The mills, now sadly derelict, remained in use for more than 120 years, until in the early 1970s, production moved a few miles away to more modern premises.

Having established his factory Stephen Moulton bought a dilapidated mansion adjoining his mill and renamed it 'The Hall'. Built in the early 17th century, The Hall had once been the finest house in Bradford-on-Avon. But when Moulton bought it, it was semi-derelict, having been used for years as a wool store. He began the work of restoration and maintenance that his descendants have continued to the present day.

Stephen Moulton's great-grandson, Dr Alex Moulton, is the present owner.

Born in 1920 and one of Britain's greatest living engineers, he has devoted his life to engineering innovation. Best known for his small-wheel bicycles and interlinked automotive suspension systems, he has also been deeply involved in aeronautical engineering and spent many years working on improvements to steam power.

Alex Moulton became a Royal Designer for Industry in 1968 and eight years later was made a Commander of the Order of the British Empire. In 1980 he was elected to the Fellowship of Engineering. He has many other qualifications and honours.

In autumn 1998, cycling historian John Pinkerton interviewed Alex Moulton at The Hall.

CHILDHOOD

John started by asking Alex about his childhood at The Hall, with the railway and river at the bottom of the garden, and the rubber factory just across the river.

Alex Moulton: I was brought up here (in Bradford-on-Avon) by my grandmother, the chatelaine of the house at that time. My mother was a widow, so really the leading person in the house was my grandmother. My grandfather had just died. My father had also died. So, I was brought up here with this very Victorian lady.

It was very much the residue, if you like, of a Victorian ethos or culture. She had lost her beloved younger son, my uncle, in the First World War, which was not that long ago before. And she had lost my father, who had died early. That was the background. And I, as the youngest grandson and youngest grandchild, spent my childhood here.

I was enormously happy and felt that I really belonged here. And it was realised that the circumstances were not at all luxurious. I mean the place was obviously lovely and luxurious, but in the times between the wars we were very tight for money.

The factory to which you refer, the rubber factory, was going through a difficult inter-war period. It was not a luxurious life but very disciplined, and I was certainly very happy here. I loved the place and that nostalgia must have been implanted in me and it carried me right through my life.

The influences included certainly the railway, the Great Western Railway, which went literally right through the bottom of the garden. And maybe during our talk we may hear a diesel train going through. It's that near!

And the other thing which influenced me, of course, was the Works to which you

refer, the rubber Works. We had a nice steam engine or steam plant in it, by Bellis & Morecambe, a well known Birmingham name in generating. We, my brother and myself, were often shown over the Works and we were very welcome. The workers always called us 'Master': "Master, come and look at this. Master, come and look at that."

And the other influence, which was important for me, was the estate carpenter. His name was Cooper, a wonderful man of some age, who interested me very much by showing me the use of tools. They were all woodworking tools. Concerning the making of things, I am sure I was influenced by him, this lovely man, Cooper.

I was also influenced, of course, by the steam. I think, even as a child (although I couldn't interpret it in this way) that I wanted to become an engineer. I was fascinated by things being made and by how these things worked. I think one basic thing about somebody who is interested in engineering is the wish to improve things. It is a natural desire, whatever these people come across and whatever they are interested in, to improve it: *homo faber* – 'man the maker'. So the steam engine was one item, and then obviously the bicycle, because that was the first mobility that you had.

John Pinkerton: When did you get your first bicycle? Can you remember that?

Alex Moulton: I'm certain that it must have been a kid's cycle. I can remember riding it about in the garden. It certainly must have been a tricycle, a kid's tricycle. I must have been between the age of four and five. I also remember going to the estate carpenter, Cooper, and I said, "Look, plane." An ambitious thing to do! And so he made, I remember very clearly, two bits of wood like that. One was meant to be a wing, one was meant to be a fuselage. And on this aeroplane I hopped about, obviously not taking off but going about.

I was very, very young then. And this is why I can remember my enormous disappointment when Bailey, the rose gardener, said to me, "An aeroplane is not made like that, it is made of canvas – canvas and wood." Obviously, he was thinking of the First World War – Sopwith aeroplanes, made of fabric. That was a great blow to me, that I was being introduced to something that wasn't right.

John Pinkerton: You were born in Stratford-on-Avon?

Alex Moulton: I was actually born in Stratford-on-Avon. I spent my childhood here in Bradford-on-Avon, but I was actually born in Stratford-on-Avon, which is my mother's home. So, as grandchildren we used to alternate our holidays, primarily here at Bradford, where we are now, but also in Stratford-on-Avon. And it was there that I have a very clear memory of being given, I think, a Hercules. It was certainly at prep school age; I went to prep school in those days at Arden House, Henley-in-Arden, which is nearby. I think I was at the age of nine back then.

I can remember one thing at my grandparent's house in Stratford: My grandfather was a family doctor, mayor for two years before my time, and then he was head of the hospital – a lovely man. And he used to arrange that, when I came back from the prep school at Arden House, there was a lovely black pig, which was a stuffed toy or a pet if you like, peering out of the door and welcoming me back. That was marking my homesickness or my longing to come home. I've just taken up, in my old age, having a cat. We certainly had a nice cat called Evelyn there. This is something which is a little bit sentimental.

Coming back to the bicycle, it was, I think, a junior Hercules. And I think it would have been in the well-known category of £3 19s 6d. It was bought in High Street and I remember going and seeing it, with its enormous rod brakes. So certainly, this was my first real bicycle. My sister and I had an enormous delight in definite little routes in the countryside around. It was like an emancipation, that mobility for the first time.

John Pinkerton: A lot of people would agree that the bicycle is 'the freedom machine'.

Alex Moulton: 'The freedom machine', absolutely! I have spoken about my delight in the bicycle and cycling. I did, by the way, make the long ride from Stratford to Bradford on that bike. I suppose I would have been 12 or 13, it might have been a full-size Hercules then, but it certainly was a very basic bicycle. It was about a 75-mile ride, it was a good ride. The impression of that, which came very deeply into my mind, was that I was alone and that I could make up my own mind. There was one grandparent at the opposite end of the journey, the other grandparent and the chauffeur here. But I was absolutely on my own.

The magic was not the local mobility, to which we referred, with my sister around Stratford, but this long venturesome journey. It struck me that I was absolutely in love with the concept of the Cyclists' Touring Club and the long distance aspect. Therefore, I became a member at a very early age. I think I even was a member of the CTC at the age of 13 and I was always looking at those lovely CTC signs. The romance of the long distance ability of the cycle moved me very much.

STEAM POWER

John next asked Alex how he became interested in steam power. Apparently, it all started at prep school.

Alex Moulton: There was an amazing thing. There was a boy there – his name, I think, was Smith – and his parents owned a fleet of Sentinel steam lorries and steam waggons. And I used to romanticise about these vehicles and study all the catalogues and see them and so on. So, there was an early imprint in one's life, which was steam. And I wasted in later life, in retrospect, a great deal of time and energy on developing an improved, more efficient steam engine. I can trace that now to an early interest in locomotives here and this kid whose father had a fleet of steam waggons.

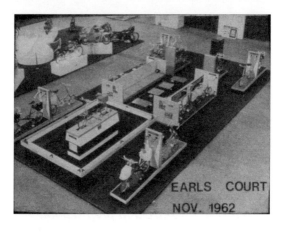

EARLS COURT
NOV. 1962

My father, who had died, and my uncle who was killed, were both at Eton at the end of the last century. It was very much a tradition that boys go to their fathers' school. But it was significant that we were, at that time, in financial pressures. It was a thing that I always very much respect, that our money which the family lived on was dividends from the factory. The factory was just a yard away, as it were – just over the fence.

It was a difficult period in the 1920s, with the slump of 1929. And therefore, we were 'hard up' as a family and very properly, my grandparents, my grandmother and my mother decided that us boys, that my brother and myself, should nonetheless go to public school but to a less expensive one. I was sent to Marlborough, which was actually in those prewar days very rigorous and very, very good. I'm extremely grateful for that education, which was the best I could possibly have had. But after that, carrying on this theme of steam, I had resolved to go to the Sentinel Waggon Works as a pupil apprentice.

Now, a pupil apprentice: in those days, the parents paid money for the boy to be trained. It was an extremely good thing. And the factory was small enough – 800 people, that sort of size, doing steam waggons – to have all functions of manufacture comprehensible. You could see the pattern shop, the machine shop and so on. One particularly interesting project that they were doing at that time involved Abner Doble, who was an American acting as a consultant there. A very interesting high tech improved steam lorry was being developed. And I would just interpose that during my school time at Marlborough I was actually making a steam car here and taking bits back to school. It had a GN chassis with a steam engine in it and I actually ran it round the grounds. I've heard tell that it exists still, which is very exciting. But I sold it before the war, got rid of it. It would be nice if I could get hold of my original first thing that I made, that steam car with GN chassis and chain drive, to put it into my museum.

So, that was a bit of the steam history. It ended up very much in a valuable pupillage at the Sentinel Waggon Works in Shrewsbury before I went to Cambridge and did my mechanical sciences.

John Pinkerton: I think it's been said that, if the effort that has been put into carburetion and ignition in the internal combustion engine had been put into steam production and then into condensation in the steam engine, we would now be running round in steam vehicles. Is that true?

Alex Moulton: That is a fascinating hypothesis, which I have heard. Up to a few years ago, I would have accepted it. But what in fact has happened is that the fundamental efficiency of the internal combustion engine is fundamentally very high and that's why the thing has survived. And the steam engine is fundamentally inefficient and one has to do all sorts of very complicated things and yet still end up with less efficient results.

What you say could have been true if it had not been for the fundamentally lower efficiency of the steam cycle compared to the combustion engine. Also, a point that you should mention is that automatic transmission (which is very much an American invention developed at General Motors) was enormously significant, because it provided the internal combustion engine with the quiet start and the acceleration.

MOTORCYCLING

As soon as he was old enough to get a licence, Alex Moulton took up motorcycling. His first machine was a Scott TT Replica. And knowing how idiosyncratic Scotts were, John asked if it was a good one.

Alex Moulton: (Laughing) I remember talking of that motorcycle. I had it back at Cambridge during my first year (we're talking before the war). I don't know if we were allowed to have it but I certainly did. We did the tuning of it in the long corridor of the engineering lab, without an exhaust system. We obviously did it out of hours. My goodness me, I can remember to this day the noise of the Scott, without silencer, being set for the Amal carburettors and running down the corridor.

John Pinkerton: Was that your only machine or did you have others?

Alex Moulton: No, I've had a lot of them. But let me just describe one little incident of that Scott motorcycle which is relevant. I was going to Malton in Yorkshire on the Scott and it must have been the winter of 1938. I was setting out from here, Bradford-on-Avon. In those times, of course, one's kit was leathers, with the Brooklands helmet. It was winter and one put newspapers around oneself to try and keep warm. I had not realised that this was a hell of a ride from Bradford-on-Avon to Malton, Yorkshire in winter.

And it began to snow when one got north and I was falling off rather often. I was actually going quite slowly but I knew that I couldn't make it because the snow was getting worse. I didn't hurt myself because I was just laying down the bike and I was going very slowly with my feet probably on the ground. I was picked up by a nice guy with a lorry, a truck. I said where I was going and he said, "Well, I live near there". And he lifted this bike on the waggon and said, "I'll take you to a farm that can possibly put you up." And I remember I had a funny helmet and this kit on, and it actually wasn't leathers, it was a black suit that I was wearing. And I had a pack on my back and I knocked on the door. And they thought I was a parachutist. They were very scared, they thought I was a German parachutist!

AERO ENGINES

So, turning to things aeronautical, how did Alex get into aeronautical engineering?

Alex Moulton: On came the war, we are talking 1939 now. At the outbreak of war, everybody of our age immediately volunteered and I, living here at Bradford, was told to go down to Bristol for recruiting. They told me, as I was an undergraduate, to go to Cambridge, which I did. So I went to Cambridge, to the recruiting board there, which they had set up at university. It was in the Senate House: there were long lines of people there. The particular man that I saw, he must have been a don, was Wing Commander Henn. I said that I wanted to go into the Air Force and he asked me if I'd had any flying experience and I said, "None at all". He said,

"What about dinghy sailing?" And I said, "I don't know anything about dinghy sailing. I paddle a canoe. But I want to go into the Air Force." He said, "But we think that if you are a dinghy sailor, you will be familiar with the feel of the air in the sails, you see?" That was the end of that conversation.

So they said, "Well, you're doing engineering, part way through the engineering course. We will put you down for the Engineering Branch of the Air Force." So I was put down for that.

At that time – we are talking of September '39 – there was no opening because everything was overloaded until April 1940. That was the date that they thought I would be needed. So I came back here to Bradford and said to myself, "Look, no way am I going to hang about with this war on till April next year. I must find out about aero-engines, as I'll be an engineering officer."

A Marlborough friend of mine, Michael Robinson, his godfather was Roy Fedden, who was a great engineer at Bristol engines. I had met with him, so he just vaguely knew that I existed, as a friend of his godson. So I went down in trepidation to Bristol, to Filton, and asked if I could possibly see him. And I was allowed to do so. I think I said to him, "Mr. Fedden, I have got three or four months before I am called up for the Engineering Branch of the Air Force. Could I work over here and get experience?"

So he pressed some buttons and said to an assistant, "Show Moulton round and bring him back when he's finished." I came back and said how enormously interested I was in almost everything I saw. He said, "Which department interests you most?" I said that I was fascinated by the single cylinder testing. Remember, the Bristol engines were radial engines, they had single cylinder cells: I said I had an enormous interest in that. So he said, "Right, you will start working for £4 a week." That sort of thing, it might have been less than that, it might have been £3 10s.

So I immediately started to work. I went in and I had digs with the parents of my friend Michael Robinson, Fedden's godson. I used to commute on a Norton. It was the single cylinder, pushrod, overhead valve, Model 18 – big, long stroke, 500cc. Rather a low spec one. I used to commute from Westbury-on-Trym to Filton every day. And I had a most enormously interesting time there, working in the test shop as a fitter. That gave me the interest in, and to some extent slight skills (I had modest skills from my period at Sentinel) in fitting and making those things. I loved all that actual practical work and the quality of those aero engines was, of course, fantastically high and marvellous. And so, I was very, very happy there and made a little bit of a contribution.

I remember one thing, a tiny thing. I was working in the Rig Shop, which is where the superchargers and carburettors were, as opposed to the Research Department. We had a problem on the Mercury engine, which was fitted in the Blenheims. There was a whole batch of Blenheims being sent to Finland and it was a problem of cold start. We were all preoccupied with getting a cold start in this en-

gine. And I suggested and made and used a little probe with a little swirl jet into the outer casing of the supercharger, which broke the fuel up to help the start. And that little device I had used in my steam car. In other words, it was something that I had developed before for the burner of the steam car. I used it in this thing and it helped.

The people I was working with and my immediate boss said, "This guy seems to be rather helpful, so we'll reserve him." In those days, front companies like aero engine people obviously had reserved occupations. So, I was put on reserved, which all the time reflected whether the person was any use. And then April passed, the date when they might have called me up for entry into the Air Force.

John Pinkerton: Do you feel any regret about that?

Alex Moulton: Yes, very much so, throughout my life. And particularly as all my friends were serving, nearly being killed all the time. That has been a great regret. On the other hand, there's a little offset of this thing. In September we had a very bad daylight raid and we were in our shelters nearby, near the works, looking out, peeping out of the shelters, to see about 80 of the Heinkel 111s with a fighter escort coming up from the Bristol Channel. It was a fairly low, very accurate daylight raid. It was a terrible destruction. And we were fortunate in our particular place. Our shelters were the oval section concrete things, which were half underground and a nearby one and more other ones had direct hits.

So, it was a terrible carnage. 200 people were killed, I think, and the consequence was that two days later, Roy Fedden, the chief engineer, summoned me.

He said Adrian Squire of Squire Cars had been killed in that raid. He was a young man, he would have been around 30, who was a project engineer on the big Centaurus engine, which was the big 18-cylinder engine. And Fedden said, "You will take his place and know everything. You will act for me, you will be my eyes and ears and the message carrier. You will have to know everything about this engine." So, I had out of that tragedy an extremely interesting war – by good fortune I was not in the shelter that was hit.

John Pinkerton: You probably did not realise at the time that the contribution that you were able to make together with the Bristol Aeroplane Company toward the war effort was far greater than what you might have been able to do working in some workshop as a fitter, somewhere out in the war zone.

Alex Moulton: Oh, indeed. I did not realise that.

John Pinkerton: Was there any animosity from the population, that you were left behind, a young chap like you.

Alex Moulton: Very interesting. No, I did

not feel or sense that at all. Rather, I was sensitive of it, in a guilt mode. But in no way was that so. Look, obviously, I was young. But there were other young ones in the shop. And the engineers in Bristol were not old. They were older than I perhaps but all modestly young.

Let me make this point about when the night raiding came into Bristol – nothing to do with the factory, which was a perfectly legitimate daylight target. About then we had the end of the Battle of Britain. It was wonderful to see the fighter defence coming in. A squadron of Hurricanes came in to Filton aerodrome and they came in to defend us, entirely by air: even the little Dove aircraft bringing the supervising person. They didn't come by road: it was a complete air movement of the squadron and the support.

The Germans tried a raid on Filton again, two or three days later. To see, looking out of our shelters, the enemy absolutely being dispersed by the RAF, it was quite moving. So therefore, we were associated at Bristol in being the people who would help "get on and make retribution".

We were providing the engines which were giving retribution for the night raids, which came very soon. So we were regarded as "get on and do it" people. And I feel that the bombing was the only thing we could do, and our engines were very important in that.

AUTOMOTIVE SUSPENSIONS

After the war, Alex Moulton had the opportunity to stay with Bristol Aero Engines. Instead, he decided to return to the family firm, Spencer Moulton.

Alex Moulton: I felt that I could, and should, contribute something to the future of that industry, of that manufactory which was primarily concerned with rubber applications to railways, dating from the nineteenth century and in which we were very successful. I felt, even at that time, that the automobile should be a target for the application of rubber suspension.

John Pinkerton: And was it your grandfather who was in the Moulton industry?

Alex Moulton: No, great-grandfather. My grandfather was the chairman of it and

my uncle who was killed in the Somme, was in it, so there were three generations and a few other uncles and so on. We owned the factory but not the firm. It was a public company, two families: Moulton and Spencer. Spencer were the consulting engineers in London. A very important point this: often when family members were killed off in a war – as the Spencers were and the Moultons were – then the remaining member, which in my case was my grandfather, who was the chairman – wasn't all that interested.

So, the company was a bit run down and not very strongly directed. I, as a young man, knocked on the door and said, "Look, could I possibly come here?" The Spencer people who were there said yes. So, I went in as a young man and first of all learned everything about it. I became assistant to the Works Manager who was a lovely man, Chrystal, and a marvellous chemist whose name was Sam Pickles. Can you imagine a nicer name? Dr Sam Pickles, who was a lovely Manchester University man, a very distinguished rubber chemist, absolutely of international standard, who by chance liked this little firm. He could have taken any job.

I was very much taken in hand by him and by the Works Manager and was really guided along. I soon realised that I must learn all about the fundamentals of rubber-to-metal bonding and the whole rubber technology. And I was allowed to set up, with the influence of the directors, the Research Department in 1946, or 1947 I think it was.

There I studied these fundamentals. At the same time I was being persuaded by these two wise man, Pickles and Chrystal. They said, "Alex, you ought to go back to Cambridge and finish your degree." So, I went back. Very much a mature student at the age of 27 in 1947 and I enormously enjoyed that return, because as a mature student I knew what I wanted to learn. I sensed my ignorance and I applied myself to it. That was a wonderful 18-month, four-term period.

John Pinkerton: Did you feel a bit sort of out of it? Ten years almost gone?

Alex Moulton: No, you know there was a whole flood of people who had served in the forces, so one had plenty of contemporaries. That was a wonderful period, because the one thing that had not changed was the university, the continuity of the whole place. And of a lot of the staff were the same. So, the continuity of the institution was still absolutely there and one's contemporaries had the wartime experience. There were youngsters and it was nice to be with one's older age group and the youngsters at the same time. So, it was a highly worthwhile period.

I came back to the factory and set about pioneering these new rubber suspension ideas, and I had a whole raft of new innovations that I had developed: the Flexitor, the cone springs, various other devices and the Hydrolastic systems. All these were born, if you like, in that study of fundamentals with a tiny team. Some of the team even carried on in my life, because they were Bradford people.

So, it was a natural continuity of a technology in which making things was fundamental.

John Pinkerton: Wasn't there a special little motor car which you were involved with?

Alex Moulton: How clever of you to remember. Yes, a thing called the Dragonfly. A friend of mine, a contemporary of mine at Bristol Aero Engines, Ian Duncan (older than I), had conceived this tiny vehicle using a Triumph Speed Twin engine with four wheels and I put rubber suspension on it. It interested Len Lord, who later became Lord Lambury. It was a forerunner of a thought of a small car, but in no way to be connected or associated with what then enormously manifested itself 10 years later. The Dragonfly interested Len Lord whom you know from the Birmingham connection. It was looked at it for a certain time and then it was scrapped. And that was that.

But the important thing is coming forward from that period to a very much more important period of the mid 1950s, when I had got to know Issigonis, the famous Alec Issigonis. I was introduced to him by David Fry, of the Fry's chocolate people. He was a great racer at Shelsley Walsh, the speed hill climb and so on. And Issigonis, of course, had this lovely thing which was called the Lightweight Special, a beautifully made thing. And a very important point on the Lightweight Special that Issigonis had made was that it had a rubber band as a springing means. It is often wrongly associated from that origin, that he, Issigonis, was interested in rubber suspensions. He was not at all. He thought it was a very bad thing and in no way to be used in a serious car. Per-

fectly all right for his little Lightweight Special. So, that is a piece of non-wisdom or a myth, if you like.

John Pinkerton: Did you ever come across a chap by the name of Bob Collier, who worked for the Norton Motor Cycle company and produced a rubber suspension for a motorcycle?

Alex Moulton: No, Greaves, of course, I knew.

John Pinkerton: No. This is Bob Collier. But he did produce motorcycles with rubber suspensions on the back, in partial conjunction with Dunlop.

Alex Moulton: Oh, did he? Would it be the MacBeth system possibly?

John Pinkerton: No, I think it was the Collier system.

Alex Moulton: No, the answer is no. But talking of motorcycles, just by the way, I did fit up a Vincent HRD, a Comet, with rubber torsion Torselastic suspension on the rear in about 1950, so I have done a motorcycle. Also I did an Enfield, because during the war Royal Enfield were down here and after the war they continued motorcycle manufacture in an old mill building here for some time. So it was easy for me to get to know the managing director, I think Young was his name. We certainly made quite a nice Enfield, with rubber suspension on the back. But let me just finish, so as to close in on the motor car suspension. I got to know Issigonis, that wonderful great genius of motorcar engineers, and we became great friends. However, he had no interest in rubber suspensions at all. I, therefore, did a very important bit of work with Jack Daniels at Cowley, who was in charge of the British Motor Corporation Research Department. We fitted up a Morris Minor entirely with rubber suspensions, took off the torsion bars (I've got the fragments still). We did a test at MIRA (Motor Industry Research Association) on the pavé. It was a thousand miles straight through without any failure at all. And it was a fact that motor cars in those times typically did 500 miles and broke down. This was a demonstration that, if you do it properly, rubber suspensions are a sound thing. And that impressed Issigonis.

Before he went back to the British Motor

Corporation, Issigonis and I worked together on the Alvis car at Coventry. The car itself was not successful but we invented between us a concept of fluid interconnection, which is the basis of Hydrolastic. When he was brought back by Len Lord to be Chief Engineer at BMC, Issigonis wanted my devices for introduction in his cars. That was the start of my very fortunate long association with Dunlop, who were to be the manufacturers, and the British Motor Corporation, who incorporated my suspension in the Issigonis-designed cars. And in my little life that was an enormous experience.

John Pinkerton: I remember that period fairly clearly because I worked in the Catering Department of the Exhibition Hall up at Longbridge. And of course we used to produce the meals that went across to what was called the Kremlin.

Alex Moulton: I absolutely remember that.

John Pinkerton: I remember all those funny little cars that used to come along. What was the one from Wolverhampton?

The Whisk or Wisp or the Fisk? And I probably saw that Dragonfly.

Alex Moulton: Hold on to that period for a moment, because you are talking about the late '50s and the beginning of the '60s. That was a tremendously dynamic, wonderful period of the British car industry. It was the very dynamic leadership of Len Lord and George Harriman, very much Longbridge, very much the Midlands, very much Dunlop, all that wonderful industrial centre, really doing something, a dynamic of world significance.

We did not feel gung-ho: we just felt that it was a natural thing to do. And that is why it's so sad now (though slightly cynically amusing) to see BMW and Volkswagen fighting with one another over Rolls-Royce on the one hand and Rover on the other. It's a rather sad thing, isn't it? In that period that you refer to we were really dynamic and potentially world dominant.

John Pinkerton: Do you see in the current situation any real future for Britain? Because we seem to have lost all of this and now they are talking almost of closing down Rover. We often have to import the manufacturing from elsewhere yet we were sort of the world leaders.

Alex Moulton: I feel – and I am obviously of the age of having had the direct experience – that the contrast is tremendous. I feel therefore that it is enormously sad what we have done. It can be clearly stated in my view: we have simply looked to the money gains. What is the money gain? It's the manipulation of money. The problems we have currently result from the folly of bankers, the greed of bankers and of the manipulators. There is no other fundamental change, and yet there has been a tremendous chaos brought about in the financial sector.

John Pinkerton: We've left it to the bean counters.

Alex Moulton: That's it, absolutely. Whereas the correct thing I very much believe is that man the maker – *homo faber* – should make things for the use of other people, for himself and other people. Sell it with a profit: it is a very important thing not to sell things with a loss – you would not survive then. But do not take

the money as a prime judgement, which absolutely happens now.

The new government takes an enormous interest in the new rich and the prime thing is, what is the money gain? Whereas we should all be considering what is the long term issue. And I am constantly crying, let's take a long-term view. It does not matter a damn whether the profit is a little low. You don't want it to be in a loss for too long but it doesn't matter about the profits. Who are the shareholders? They are only institutions! The long term issue is the key.

John Pinkerton: You have been involved with Issigonis and you obviously have a great regard for him. And jointly, you have achieved some outstanding successes.

Alex Moulton: Well it was wonderfully synergetic. I mean he was an older man, about 15 years older than I. But I learned so tremendously from him, his methods. Of course, he was not at all didactic. I am interested in trying to explain things to younger people. And I am very much concerned with working with Bath University and the Smallpeice Trust doing engineering guides – one page, one item – in the most direct way of solving particular technical problems of engineering. I do it every week with a very nice representative from Bath University and it will be published by this charitable trust.

I want to concentrate on the heritage aspect of this important house, but also on engineering education, because I deeply believe in the significance of engineering – *homo faber* – and I deeply believe that our educational methods are not so direct now. Therefore, anything that I can do gives me a deep satisfaction. Issigonis was very much a prima donna, which he quite properly should have been. One simply learns from him, from his works: if you examine his creations one can learn from him. And I was fortunate enough to be directly associated with him and learned a lot directly.

John Pinkerton: Isn't he very much like a lot of wonderful designers and wonderful inventors who tend to be somewhat introvert? They can do the job and then they've done it. You, I think, are quite different. You want to tell people how you design.

Alex Moulton: You are absolutely right in that definition and that differentiation.

I think that I am driven by a desire to understand and to check, check, check, check, if what I am doing is reasonable. Therefore, if you go through that process, part of the process is to explain it.

John Pinkerton: Even if you only explain it to yourself. And once you have done that you can explain it to someone else.

Alex Moulton: Yes, absolutely.

John Pinkerton: So, when Issigonis produced the suspensions for the Mini and so forth, you were still at that time associated with the factory.

Alex Moulton: Yes, we sold the factory just about that period. I as a young director of the Spencer Moulton factory could see that from a synergy point of view it was the wrong size. So, I was instrumental in getting a buy-out of our neighbouring and much bigger factory, Avon Rubber, which happened at that time, 1956. And that was an enormously successful thing. It was a real merger if you like, in which there was no strong ego of the Spencer family, because I went off with my own separate business, Moulton Developments and the bicycles and so on.

Therefore, it was a proper takeover. It was all very local. The Avon directors recognised the strength of the Bradford people and, coming forward goodness knows how many years, the factory at the bottom of the garden here is no longer a factory but the operation has been transferred to a greenfield site at Chippenham. And the key personnel are from Bradford. So there is an enormously successful tradition because of the father and son continuity from here. And by the way, I am very lucky to be a technical consultant to

Avon after all those years. So the answer is, we sold out sensibly for synergy for the continuity of the rubber manufacturing business and I myself started in these stables, doing this research for the British Motor Corporation. The products would be made by Dunlop because Avon was too small.

THE MOULTON
BICYCLE

*What got Alex interested
in developing the bicycle?*

Alex Moulton: The launch of the Mini was
in 1959. The impact of that on the world
press was, of course, tremendous.

Extract from BMC advertising film: "Over
rough track to see how the new rubber
suspension stands up to the shock. Tests
that go on and on! These motoring cor-
respondents took the new cars and saw
for themselves how the rubber suspension and the new engine position gave
them a spacious car with such an ability
to hold the road that they were a zippy joy
to drive. These new cars are sensational.
Some baby, loaded, the most sensational
car ever made here."

Alex Moulton: We were all very buoyed up, because we were taking part in this and then it moved very sweetly on to the Hydrolastic on the 1100, which was going to come out in 1962. All had been prepared. And I was buoyed up by having seen success in which one was swept along as part of it – in no way responsible. It was entirely Len Lord driving from the top and Issigonis with his brilliance and Dunlop, who were wonderful people.

One knew what success was like and what you do for success. You take every care – a very fundamental thing. Like we did –

if you think back – the shadow factories during the war, it was the same sort of attitude. It was the continuity of that winning attitude if you like.

So therefore it was with the petrol rationing, you know, the Suez Crisis of 1956. We were all tending to use smaller cars, because we were rationed with petrol. I then simply for the first time tried a nice lightweight bicycle. One of my friends from the Bristol Aeroplane Company days had a Hetchins and I bought this Hetchins and I was absolutely overwhelmed by it. It's in my museum now, how wonderful it is. I had no idea what a lovely bicycle, a fine lightweight bicycle, was. I thought, my goodness me, what a fantastic thing. I just could not imagine it. So I was moved to say that I would like to improve this.

John Pinkerton: So you had this delightful experience of riding a hand-built beautifully crafted lightweight bicycle. It must have been a tremendous change from your 'gaspipe special' Hercules. And that had a real sort of impact on you?

Alex Moulton: A profound effect. You see, I had not bicycled really since the Cambridge days – nothing – and the very kid days, making a long journey as a child. And so this was a revelation – and a revelation of joy. A very important point: you cannot create anything out of hatred or disinterest. You must be driven by a joyful feeling. And I was also joyous by having seen how technical success is achieved, by the Mini and so on.

So, I was determined to do it in an analogous way. In other words, I would study fundamentals and then decide what to do. I rode a recumbent, an FH Grubb, which I imagine was copied from the French prototype. I was questioning the riding position. While riding it I immediately found the thing that is rather different to what people think nowadays, the better acceleration by being able to get a purchase from your back. The other point was that I immediately noticed the difficulty of holding the femur – the weight, the mass of this top piece of leg – up, because I hadn't considered putting clips or pads on. That was a disadvantage. And the other disadvantage – which sounds rather funny to-day – one got splashed by lorries.

But above all, and I say this deep truth too today, it was unstable. In other words, the natural instability was difficulty in balancing of the very low centre of gravity and the low inertia, of which the complete opposite is a High Wheeler, of course, where you could maintain – you know that better than anyone – stability at a tiny speed. The deficiency of that really alarmed me.

Coming across the years – because I have been impressed by many people who made recumbent Moultons – I had a look at that and did a long study, two years ago. I did a study with students from Bath University as an exercise. I said, am I right in still disregarding the recumbent version? And I did a test two years ago – a very simple test – with a well-adjusted young rider. In other words, I was testing the control feedback loop between the eye and the hand, which was well established in this young man. He was to set off at the top of the hill, run down freewheeling through the archway and through the stable yard. And when he was in flight, I would throw a leaf onto the flat surface under the archway and the charge was to avoid it. Now, he couldn't take a prior aim, because he didn't know where I was going to put the thing down.

It more than confirmed my belief that, in comparison particularly with my Moulton bicycle, which is extremely 'dartable' and controllable, that he was in no way able to make such quick changes of direction, which is a fundamental feature of vehicle stability. So I confirmed 30 or 40 years after the first experiment that I did not want to go the recumbent way.

So, then coming back to that early period in the late '50s, I looked at the rest of the thing, as I had no complaints about components, because everybody over the years had been developing components. And we, of course, in this country had a wonderful component industry, such as Williams cranks. Apart from the derailleur, by and large everything of the componentry of the bicycle had come from England. Is that a reasonable statement?

John Pinkerton: In general terms. Even the derailleur: there was a derailleur on the Whippet of, I think, 1896. A two-speed but it was a derailleur.

Alex Moulton: It is a general truth that our wonderful Midlands produced the infrastructure of cycle components. And I had no complaints with that, the way they had gone. They had done well: they had developed it well. People did not say how bad a brake was or how bad a saddle was. But what struck me immediately (having settled the riding position, the classical riding position) I said to myself, why are the wheels so big?

Look, I was moved by the enormous importance of the size of the wheel on the Mini, on the whole arrangement and space occupation and architecture of the Mini. It was influenced tremendously by the tiny wheels and they were tiny wheels in those days. I naturally had no fear in questioning that, working of course with Dunlop, with whom I was very closely associated with the car suspension work. Dunlop in those days were wonderful. They were really enthusiastic about doing tests. So all the work on rolling resistance was done with Dunlop making new tyres, making new moulds and so on.

And so it was the small wheel which was really the significant venture, with high-pressure tyres and naturally suspension, because I just thought that it was an unusual thing for any human-bearing vehicle not to have suspension. And then the open frame and other features followed. So the approach was a fundamental approach.

What I did not like I altered in the manner that came out with the first Moulton bikes. That was, if you like, the creative origin of the Moulton bicycle in the form of very good prototypes, because we had skills in our workshop for developing suspension and we had this lovely workshop. From that, by the way, right across the years – 40 years or so – I maintained that the key help in engineering creativity is prototyping, the skilled men to translate an idea from the mind, delineated on paper by drawing or some sort of a sketch, into hardware for testing. It was the key thing I learned in those days, and I maintain it absolutely.

John Pinkerton: We are going right back to The Hall carpenter idea – the man with the skill in his hands. I presume then that you tried to sell the idea, because I don't think you wanted to be a manufacturer.

Alex Moulton: You could not be more right. May we just talk on that, because you remember that I had come from a family of rubber manufacturers, had seen if you like the very much larger scale which is necessary for the manufacturing of suspensions, needing Dunlop (a much larger firm – Dunlop of Coventry, that was) and I had thought to myself that the natural thing to do is to license the Midland skills to manufacture this thing. I am creating it.

Who to go to? I had, through Issigonis, an introduction to TI – Tube Investments, Birmingham. I had an introduction to one of the original directors of TI. I remember I had a Lancia Aurelia car and one of the prototype bicycle models, which is now in my museum. I was delineating this new form and Issigonis took me to this guy. I remember putting the model of this thing on the roof, because it was quite low when you were looking at this thing. And the TI man said, "Don't go to us, TI Phillips, go to Raleigh," who were in fact, at that time, their competitor, "and go and see George Wilson, who is the chairman." And so I did.

It ended up with him coming here. He came down in a big Jaguar with a standard bicycle. I said, "Bring with you a good standard bicycle". I had a rider here with me and he demonstrated the bicycle up and down, the Moulton bicycle – one

of these nice prototypes – up and down
the hill from Staverton, which is quite a
steep hill. George Wilson, comparing the
two bicycles, took the standard one out of
the boot (he had this Mk 10 Jaguar with
a great big boot) and rode about on the
Moulton a bit. I don't think he struggled
up the hill, but he just watched obviously
our rider Woodburn going up and down.

There was no doubt that the Moulton bicycle was enormously comparable with
this Raleigh bicycle. I remember George
Wilson saying, sitting on the wall over-
looking the river at Staverton, "Gosh, I
want a fag after all of this." So he took
out a Woodbine – he was one from that
generation – and I got that message very
clearly. He said, in a moment that was
very significant in history, "Look, do you
mind if I take back this prototype." I said
it's enormously confidential. He said that
it would be absolutely confidential. He
said, "I've got a new man coming in from
Hoover, Leslie Roberts, with whom I have
to discuss it. I want to discuss it in confidence with him."

That was the start, if you like, of two

years of consideration. And what happened in those two years? I was going up there to describe the bicycle and collaborating with them on any changes to make it more suitable. I remember one particularly important thing in that two-year period, when the bicycle was on offer. At that time we were thoroughly discussing it with the top people and I was in the boardroom at Raleigh with George Wilson. I went outside in the corridor to get something and I saw the sales manager, a great big man, Harrison, dragging my prototype bicycle sideways. His face was all inflamed and I thought, "Why is he dragging it sideways in a very strange way?" And I saw him trying to indicate that my bicycle was flexible. Actually it was enormously more rigid than anything he made. It was my first introduction to somebody wanting to destroy the stupid thing, there was hatred in it. So that was negative.

I was then taken out by Harrison and Davis, the Phillips Sales Manager. By that time TI and Raleigh were coming together in an amalgamation. And these

were the two sales bosses that took me out to dinner. In the bar I remember one of them saying, "What do you want to do, Alex Moulton?" And I said, "I want to license you to make my bicycle, so that you can sell it." But they thought that I was up to some trick, to want to try and take their jobs or take over something. There was an amazing attitude: you know, it was a corporate attitude. Here was myself, a man from outside who was familiar with the boss, Wilson, and had not come up through the ordinary way, with this amazing thing that was rather well-made. There was no question of them questioning it, because it was a beautiful product, as you can see now in the museum. "What is he up to?" they thought. And here I was, as I said, I simply wanted to license the bike.

So I saw then, as my naivety was loosening a bit, that it was unlikely they would do anything. And then of course they were thoroughly preoccupied with the takeover situation, with the jostling of the position of who was going to be boss. That was a takeover, of course, by TI as you know, and actually George Wilson did succeed in becoming the boss for a while. But properly they were concerned with that and then my product was put to one side.

Another very important point to record

concerns the marketeers. Not only the inside people of Raleigh and TI, but also very big dealers were brought in. They said, "This is unmarketable, you cannot sell it at all. Don't do anything like this." It was an enormously typical negative reaction of a corporation.

John Pinkerton: Call it the RC factor – Resistance to Change.

Alex Moulton: Yes, absolutely, and also NIH – Not Invented Here. So, what I did in 1960 or 1961, as I often do, was to question myself: "Am I right in my conviction? Is this thing unmarketable?" Or on the contrary: "Is it marketable?"

So I did a tiny market survey in Ayrshire, on the west coast of Scotland, in a very good bicycle shop in Irvine. I got a youngish market survey company in London. I said, "I can't afford any significant money, but do the intelligent thing – take it far, far away. Do this market survey in a shop far away, where nobody would have heard of Moulton."

And this guy in the small market survey operation simply set up two prototype bicycles which we had – these beautiful prototypes – one in the shop, standing up there and one just standing by. And when the customers came in, buying ordinary things such as torch batteries, they were asked, "By the way, this bicycle, would you ever consider buying it? The price of it would be 20 per cent more than that of a very minimum one." In those days it must have been £28. And if there was a positive remark or a positive interest, then they could take a test ride round with the other one as well. There was an overwhelming affirmation that it was marketable. Over 50% of people said yes, even at that price, a little bit more expensive than the cheapest, they would buy it!

John Pinkerton: But you were getting more for your money, because you were getting suspension and a unisex design.

Alex Moulton: Yes, carrying capacity, universal frame and all that. Exactly: it was a tremendously important vindication: don't take the expert's point of view, actually go to the public! But of course you've got to do it very professionally. And don't do it for oneself, because I was too enthusiastic about it, too persuasive.

But this deadpan little man did the thing so well. That convinced me. So then I decided to do it myself.

John Pinkerton: Tom Northey's (marketing guru and inventor of the Pirelli calendar) number one rule: Never overlook the obvious! You went to the people and asked them the question without any bias at all. And you got the right answer, directly.

Alex Moulton: Exactly!

John Pinkerton: So that must have fired your enthusiasm again. What did you do then?

Alex Moulton: I then said I'm going to make it myself. So, back to fundamentals. Let's build a factory on our estate just up there. So I built this factory on what we used to call the donkey field, because that's where we used to have a donkey – a wretched creature that I was terrified of as a child! We have a donkey shed there, a very nice stone building which now has an engine inside it. Then the factory was being built – 5000 square feet. I had no hesitation: I knew it was the right thing to do, to do it oneself.

And at that time, I don't know how, David Duffield entered my life. I can't exactly remember how it happened, he must tell you. We became, as we remain to this very day, great friends. He is a brilliant, marvellous man; I mean, a proper cyclist and a very extrovert, nice, open-minded man. He was my marketing man. He was quite young but had experience in, I think, Phillips cycles at that time. I think he was in Phillips to start with: he has had his whole life in the cycle world, and is now a famous cycle race commentator. I consult him now for his deep knowledge of cycling.

And how neatly the product was marketed, in the sense of the advertising. We introduced this tiny little silhouette, and all that good promotion that went on at that time. We spent quite a bit of money on it and then had a major launch at the 1962 Earls Court show – very big stand and we were inundated. I remember that Macmillan, the prime minister, came. And Birmingham were with us, which was wonderful, because we were associated with the success of the BMC Mini

just before, and it was the 1100 was coming out then – more innovation, the Hydrolastic suspension and so on.

We were carried along by this wonderful support from the Midlands. And the most amazing thing, which was very unlikely, was that on this big stand we were getting overwhelmed by people, clawing at us, foreign people as well, saying, "We want to buy this bicycle." And our little factory was not made: it was just being built at the time. I remember ordering from the Earls Court show, "Double the size", so that the 5000 square feet was extended up to 10,000.

This next part you won't believe but it's true. Sidney Wheeler, who was the secretary and the director of the Austin Motor Company and very close to the board, Len Lord and George Harriman, telephoned down to me on the stand and said, "Don't hesitate to take orders. We will make it for you." Isn't that amazing: the British Motor Corporation offering to make it for me!

And then just touching on a very relevant point, David Duffield insisted with my immediate support that we must prove the principle by a sporting triumph. So there was the tremendous race that he fixed up at The Butts stadium, Coventry – the Leicester team versus the Coventry team. Mick Ives was on that ride. And it was a wonderful thing that we had made pursuit bikes. They just soared away: whatever team was on our bicycles in the pursuit just won. We vindicated it as a racing machine, it was very significant.

Then the ridiculous attempt to do a road record in the middle of the winter, which was a damn fool thing to do. This was the Cardiff to London record, so it was a downwind run. It was John Woodburn, of course, who did it. I remember it vividly; if you are following a road record attempt in a car, you have to go very fast and the regulations are that you have to keep a distance. And if you pass the rider, you have to be very careful. I was in my Ferrari 250 Coupé, I think, and I remember I was tremendously moved by the progress of this racing cyclist on a long road. It was an amazing progression.

And then when we reached Marble Arch

I had to leave the car and hurry to Marble Arch to meet him. We went out and had supper together. So that was memorable, Woodburn breaking this record. To make a very radical innovation and presume to offer it right across the board to the public, you must jolly well prove that you are right. That's where, of course, all small-wheelers that came later were completely opposite to my desire to make a totally better bicycle; they were just shopping bicycles.

That's why in later years we did speed records – the streamlining in America and so on. Quite understandably, we weren't allowed to compete in races, because it was very apparent from the experience on The Butts track and some road racing, that with our small wheel you can enjoy tighter drafting; and it's disturbing for the rest of the riders in a peloton to have this funny little machine rushing through. That's understandable. So you simply say, forget that, let's go for a record: if you are faster, then beat us.

John Pinkerton: The rest of the Moulton story has been very well covered by that well-known book *The Moulton Bicycle*. Your business was perhaps a bit too successful for you and you needed to look for someone else to take it over.

Alex Moulton: No, that's a kind inference but it's not so. The actual fact was that Raleigh were furious. Do remember Raleigh was TI and Lord Plowden, who was the chairman of TI, used to say to his Raleigh members round the board (I suppose to Leslie Roberts): "I opened the newspaper. What is all this about Moulton?"

And apparently in the *Daily Express* there was more coverage of a Moulton bicycle than of the launch of the Ford Cortina. I mean, the publicity of our damn thing was ridiculous. That was David Duffield, who was very much strongly competent in his charisma, getting this publicity and promotion. And so in the end, I suppose, Raleigh remembered they had been taken over by TI, and I suppose that they were really incited by this thing and said to themselves, "We'll show our chairman that we were quite right in turning it down."

And so, they had been working for some long time on the RSW16, and they launched it in the year '65, I think, with tremendous publicity. They copied everything, it was very extraordinary the way they did it, except of course for our patented suspension and high pressure tyres (they had no suspension and fat, low pressure tyres). An example: David Duffield had put the bicycle going about on reps' car on the roof, standing up in the vertical way, whereas at that time ordinary bicycles were put upside down. So you could ride along seeing Moulton bicycles on Minivans. And so Raleigh did the same thing!

So, Raleigh was simply 'in the manner of' the Moulton. And of course, the publicity and the advertising were enormous, and it was said that they had spent a quarter of a million pounds on the whole promotion. And the public and the dealers were thankful and said, "Oh, how wonderful, we've got a proper firm making these things." Except that Raleigh was a little bit cheaper and much better finished, particularly compared with our ones made by

BMC in Kirkby, Lancashire, the quality of which was not very good. So, it was inevitable that we would suffer from the market having been lured by the Moulton into this sensible new form of bicycle, then into this other one made by Raleigh. It had one or two features which were better: it had a lower step-over height which women found we had made a little bit high, and so our market was being attacked.

And I therefore was sensible. I haven't been sensible many times in my commercial life but certainly I was sensible here. I said, "Hey Alex, don't fight this lot. Do a conciliation if possible but don't fight them, because they are ten or a hundred times bigger; they'll burn you out." And so I was talking to Leslie Roberts, because he was wanting the Moulton all the time. In the end it was a distress sale really, because we were getting into a loss-making situation and this is a danger. If you are in a big way, if you tip into a loss situation, it's big money. Remember, it was my money in it, I never had any external funding with any of my enterprises. Therefore now, with prudence, I keep small. So if bad things happen, then one can go through bad periods: you simply say, well, provided the projections of what you are going to do are alright, see it out. But if you are in a big way, and it is the small money of a single individual, hey ho, you can end up in bankruptcy in the end: that's all, at the end. So I was sensible conciliating to Raleigh with an sell-out.

They came down here and they photographed me. Raleigh-Moulton: we go together with this designer Moulton, with the Raleigh-Moulton. And they were celebrating. So we were photographed on this terrace here. And I did say, "Look, Leslie Roberts, don't stand as if you own the house, because you don't!"

John Pinkerton: (Laughing) You've only bought the bicycle!

Alex Moulton: Therefore, we moved to stopping bicycle manufacture and Raleigh went on with it. Now, in their bosoms I don't think they had really any intention of going on with the bicycle. But they certainly had the intention of collaboration with me and they funded an important thing, which was a little research programme, which I led entirely. We did some very fundamental research, all dossiered away now, of the power required to drive the bicycle and so on – all those fundamental points.

In the early '70s I was confirming fundamental truths, asking was I right in certain things? And these deep things I have kept: I published them at the Royal Institution Discourse in 1973. It was a valuable period. But Raleigh had no intention of doing a combined thing at all. And then they had their own history and so on.

John Pinkerton: After the tie-up with

Raleigh you were disappointed that they let you down, I suppose. You could have retired and sat on the front lawn and just enjoyed looking at the world. But you did not, you've got this drive: you wanted an even better bicycle than the Moulton?

Alex Moulton: Yes, absolutely so. The answer is, I could have done, indeed. Let's make a point: I own a lovely property, which I've conserved over the years, but I am not a wealthy person. But I certainly could have done what you say. Leaving that quite apart, I shall be driven to my dying days to pursue a number of things which I am quite consistent on: pursuing the improvement of bicycle, and the spin-off, which I touched on before, of engineering education, because I am interested in the broader sense of that.

I thought in '83, many years after having sold out to Raleigh, I'd have another go. In the meantime, do remember, that I enjoyed, and they had funded, this basic research which gave me confidence and showed that my principal theme was quite correct, with small wheels, high pressure tyres, etc. Therefore, I thought I'd have another go and would do it again but in such a small way that I shall not annoy anybody. Nobody gives a damn what I do today.

And therefore I started producing, as you know in 1983, the so-called AM series, pursuing exactly what you say – a desire to make a further improvement in my theme. But all the time, in doing a new innovation, to start one does as best one can. And then you say to yourself, "Can I do it more cost-effectively?"

I invented then, which is worth looking at, the hairpin tubes. The tubes aren't codsmouthed, which is so expensive and a beautiful way of doing it. In joining they are bent round the head and seat tubes and the extra length was just lost in the centre pin. It was a relatively low-cost method of making it. And so, having made that, David Duffield and I took it to Raleigh in 1988, and asked, "Would you like to make this thing under licence?" They said, "No, thank you." They were different people by then, but they were being polite.

Then we took it to America. There was a Japanese firm that was interested and our importers, Angle Lake, thought they might be interested in this thing. But they were too difficult to deal with, so it didn't happen. Again licensing, you see. So, I decided to see if anybody in England was interested, so we had Dick Pashley down, and he said that his boys were interested, the answer was yes.

So that was the ATB, the All Terrain Bicycle, which was what we were starting with, degraded a bit the tubing, which is perfectly alright, to become the APB – All Purpose Bicycle. And it's been a very nice arrangement which goes on forever, we hope, with further improvements, made under licence with our tools at Pashley.

John Pinkerton: Without going too deeply into the two types of bicycle that you've made, they are beautifully covered in Tony Hadland's two books (see p. 94). It must be quite unusual for a single item like one make of bicycle to have a book written about it: but two books are even better!

HOBBIES AND LIFESTYLE

Let's talk about your spare time, now. I mean, do you have time for hobbies?

Alex Moulton: I am very interested in the analogue of cycling, which is kayaking. I've got lots of kayaks, because they are relatively inexpensive. Only a fortnight ago, I was paddling a Klepper canoe at Plockton in the west of Scotland. So, that is certainly so.

Then on the estate here that I've got, we have a little sporting shoot. I do that really for a social point of view: it is very nice social entertaining. The prime aspect is that it enables one to constantly conserve it, replant the Wood and generally look after the place – that is the interest of having an estate. You want possibly a creational activity in order that you actively

Alex Moulton   *Steph. Llewellyn 2002*

are conserving it. So, those two things are very active.

And apart from that, I read very much and do recreational things –walking on the estate and hiking, apart from bicycling.

John Pinkerton: You haven't talked about photography, but you are obviously interested in photography because you've got quite a lot of cameras. The whole place is festooned with photographs: some are yours obviously, and some aren't.

Alex Moulton: I am fitting up for one's old age a darkroom down on this ground floor. So, I shall practise developing and printing. My favourite thing is the little Minox, which I find attractive and analogous to the Moulton bicycle. It's a unique solution, the Minox sub-miniature camera, and I admire them very much and use them.

John Pinkerton: You said earlier on that the bicycle meant freedom to you, 'the freedom machine', which is about travel. You've travelled a lot. Do you like travel?

Alex Moulton: No, I have travelled, obviously. I've been to Africa and to America. But I am not driven at all by travelling for the sake of travelling. I always travel for a particular purpose, which is generally connected with one's interest. Obviously, the interests of my bicycle business will take me to foreign places. But I am not primarily a traveller. I am much happier in knowing a piece of local country more and more in detail. I am very fortunate to be born and live in England.

John Pinkerton: In that limited amount of travelling that you experienced, are there any particular modes of travel that you like or dislike? I mean, do you dislike flying?

Alex Moulton: Well, my greatest ambition, which will probably never be realised, is always without any guilt to travel first class, because I find travelling on aeroplanes enormously uncomfortable.

John Pinkerton: How is your daily timetable set?

Alex Moulton: Routined, very straightforward and very rigid. I wake with the alarm clock at half-past-six, put on Radio 3 (the BBC classical music network), get up, go down and open the house up, make tea and go back to my working room, which is beside my bedroom, where I do my work. And I have at least two hours of absolutely uninterrupted best state of mind on whatever work it is that I am doing.

After breakfast I do the executive thing of telephoning and so on till about half-past-nine, then go walking through the garden and then down to the factory to spend a little time there. Always taking an interest in the people and reacting to them.

Then I come back for a light lunch and very often there are people coming. Afterwards I often like to try, if I can, to have a ten-minute snooze – I think that is a very important part of recreation. And then I do something more in the afternoon.

And then in the evening I watch the news programmes, and often listen to a concert. Then I have a self-help dinner and normally go to bed between half-past ten and eleven. So, that's the sequence I run entirely. I think it's very important, especially that first two hours.

John Pinkerton: Are you a hard taskmaster?

Alex Moulton: Probably.

John Pinkerton: Do you think you could have produced what you produced without being so?

Alex Moulton: No! (Laughing)

John Pinkerton: Are you able to suffer fools gladly?

Alex Moulton: Not for very long.

John Pinkerton: Well, you've suffered me for most of the afternoon.

Alex Moulton: You have been a delightful conversationalist. Let's put it that way.

John Pinkerton: Looking back over your 50 years or so of experience of

British industry, how do you really see it today? Is there a way forward for us or are we going to be just a nation of bean counters?

Alex Moulton: Hopefully not. The great hope is for encouraging startups. There are young people about who want to make things and we very much encourage work experience in our own little shop. There is one man that I enormously admire, David McMurtry, who set up his Renishaw metrology business, which is an example of everything that is excellent in a privately owned business. So, it can be done, but the fear of the opposite is very great.

John Pinkerton: Looking back over your life, what's been the most satisfying thing that you have done?

Alex Moulton: In retrospect, the original Moulton bicycle, joined with the Hydrolastic interconnected suspension, which is tremendously necessary for small cars but which everybody has abandoned. Certainly those two things, but primarily the first generation Moulton bicycle.

John Pinkerton: Now, your New Series

machine, which I personally have said is as far ahead of the AM as the AM was ahead of the Moulton, do you see that it's the end?

Alex Moulton: No, certainly not. I very actively have started a year or so ago on the further evolution of it. Always, as I say, in the second phase, working towards a lower cost version. That is very much in my mind.

John Pinkerton: Are there any regrets in your life?

Alex Moulton: Yes, I suppose that I haven't made a great fortune out of these things. I regret that a bit. But I don't think I'm the right sort of person to have done it.

John Pinkerton: What would you have done with a fortune?

Alex Moulton: Nothing at all! (Laughing)

John Pinkerton: Well, I have had a fascinating afternoon. It has been a delight to meet you again and to find out even more about you. I am sure that there is another hour's conversation that we could have. I would sincerely thank you not only for the conversation this afternoon but really and sincerely thank you for your contribution to cycling and cycle history. It has meant a great deal to a great many people, sometimes directly, sometimes indirectly. And I do thank you for that.

Alex Moulton: I deeply appreciate that, deeply appreciate it indeed. And I have enjoyed this afternoon talking to you very much.

Hadland, Tony, *The Moulton Bicycle,* second edition, Pinkerton/Hadland, Reading, 1982

Hadland, Tony, *The Spaceframe Moulton*, 1994